You *Can't* Manage Time

But You *Can* Manage Many Priorities

By Susan de la Vergne

Alder Business Publishing

© 2007 Susan de la Vergne

To order this title, use order form on back pages.

You *Can't* Manage Time
Printed in the U.S.A. by

ALDER BUSINESS PUBLISHING
15800 SW Boones Ferry Road
Lake Oswego, Oregon 97035
www.alderbusiness.com

Library of Congress Control Number: 2007900569

ISBN 978-0-9792987-0-7

■○■○■○■○■

Many thanks to Vincent, Gary H., Mike and Shaz for time, energy, enthusiasm, encouragement and careful editorial help.

I haven't—and probably can't—say "Thank you" often enough to Gary E.

■○■○■○■○■

Contents

Introduction

Why You *Can't* Manage Time

Given the number of courses, tools and techniques on the market to solve the problem of "Time Management," you'd think that *time* was something you could actually *manage!*

But, despite what those courses and consultants are suggesting, there's really no such thing as *time management*. What would it mean, anyway, to "manage time"? After all, you can't extend it, you can't make it last longer, and you can't shorten it. You can't save it up to use later when you need it or spend it faster than it comes to you. No, there is nothing you can do with time except live in it.

Time goes on with or without you. It proceeds at exactly the same pace, every day, week, month, year, decade. It never slows down, never speeds up, and it won't respond to you no matter how you try to manage it.

So if you can't manage time, then what do you manage? You manage tasks. You manage communication. You manage creativity. You manage energy. You manage your outlook. You manage yourself.

That's what we're going to talk about here: Optimizing your energy and creativity while deepening your understanding of the circumstances that put you in the middle of managing many, often conflicting, priorities.

■○■○■○■

Chapter 1

The Problem and the Myths

Your life on any given day is a long to-do list of conflicting priorities. It may even seem to you as if you don't control many of the things on your list. They just come at you. Does this sound familiar?

"Everything is due at the same time!"
"I'm double-booked at 9:00."
"I'm triple-booked at 2:00!"
"No one understands. . . ."
"Everyone needs me NOW."
"My boss needs me."
"My peers need me."

"Everything's always changing around here. I can't keep up and it's making me crazy!"

Here's my list of things to do, just for this week alone!

This Week:
Attend Alpha Project status meeting
Attend staff meeting
Complete Beta Project deliverable—overdue!
Catch up on emails—I'm way behind!
Update problem documentation
Call back angry customer(s)
Lunch meeting with new Project Manager
Learn new switch interface technology
Write status report
Read industry journal article
Attend company picnic committee meeting
Coach soccer Wednesday night
Keep up with other personal obligations
    (spouse, kids, dog, etc.)

Pressures on You

There's no doubt managing many priorities is powerfully challenging. As if your list of things to do wasn't long enough, there are all these other forces at work. Think about what you're facing:

✔ **Management Expectations**—They say "Get it done!" and they hand you a deadline.

✔ **Peers Need You**—You've been working here awhile, and people know you know

your way around. You know the products, you know the routine. They come to
you: "Please help me with this!" How can
you say "No" to that?

✔ **Customers Need You**—And customers
are impatient. Have you ever had a customer call with a problem and say "Oh,
no hurry, just get around to it whenever"? Probably not. They want you now!

✔ **Several Assignments Need You**—At
any given time, you may be on three, five,
a dozen assignments. They're all hot.
They all want a piece of you.

✔ **Prioritization is Lacking**—Manage-
ment/leadership hasn't clarified what's
first, second, third, so every project as-
sumes it's the most important one.

✔ **Invisibility of the Problem**—You know
what's coming at you, what you're trying
to accomplish, but no one else can see
how much you have to do and how urgent it all is. Even your status report—
when you remember to write it—doesn't
describe your situation very well.

✔ **Unpredictability**—Emergencies arise
and have to be addressed. That's a fact of
life. Sometimes—maybe often—they leap
to the front of the line, knocking behind
them all the work you were already
doing, and of course, you can't predict an
emergency. (If you could, it wouldn't be
one.)

✔ **You Lack Control**—All these forces seem
to add up to one thing: You have no control over what comes at you, what order it
gets done in, how long it takes. If only you
did, life would certainly be better.

That's just what affects you personally. Let's
talk about some other aspects of this whole
overload problem that get everyone in trouble!

Pressures of Business

Business applies its own set of pressures—
like pace, for example, driven in part by competition. Competition waits for no one! While
you're sitting over here admiring the mouse-
trap you've just designed, someone down the
street is improving on the design. Before you
get yours to market, his will be snapping shut
in the basements of homes all over America.
That's the kind of imagination that fuels the
competitive spirit and makes us work harder
and harder to beat out the other guy.

Besides competition, let's not forget the pressure of short-term results. Today's business
paradigm is all about quarterly results. They
have to be better and better every quarter.
Squeeze out the excess cost, crank up more
revenue, and turn that crank faster and
harder every quarter. The alternative, we fear,
is premature demise.

Talk about pressure!

Cultural Pressures

In case you thought that was plenty to worry about, here's one more: The cultural phenomenon of living in a perpetual state of being "busy." We're all supposed to be "so busy!" When you ask someone, "How are you?" don't they usually answer "Oh, I'm so busy"? Sometimes they go on and on.

"My boss just quit and there's no one in charge, so I'm doing more than my job. That project we started last March is in a critical phase now, and meantime my son just got over the flu, and now my wife has it, but at least he got better in time for his soccer tournament which starts tomorrow, so there goes the weekend, and —"

I mean really.

Have you ever asked someone how they are and they said, "Oh I'm just waiting around for something to do, you know my life is so empty and boring and so I'm all caught up on everything. . . ."

Not likely in today's world! It's just how things are. We're busy people responding to external pressures and cultural expectations, trying to keep order in the midst of chaos. But are we going about it the right way?

The Five Myths of "Time Management"

Myth #1
Getting a lot done is easy:
Just multi-task!

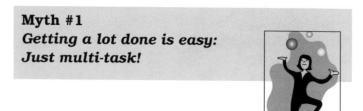

Many people pride themselves on handling lots of things at once, on being "multi-taskers." Multi-tasking is cool, it's expected. No one does one thing at a time anymore. That's so yesterday.

Sounds good, but in fact human beings aren't capable of multi-tasking. That's not to say we can't ride an exercise bike and read a magazine at the same time. Our legs and arms can operate somewhat independently of our brains. But when it comes to knowledge work, using our conscious brains, our brains can't actually do two cognitive tasks at once. We do one, then another, then another. We move from one task to another. We may even move quickly from one task to another. But what we call multitasking is actually time-slicing. In other words, we do one thing for a second and a different thing in the next second. We may think we're doing two things in the same second, but our conscious brains don't work that way.

The term "multi-tasking" came along a couple of decades ago when computers became popular. Computers can multi-task because they have parallel processors—more than one brain at a time working on more than one thing at a time. Hey, if computers can do it, so can we, right?

But alas not. When we say we're "multi-tasking," we're actually just time slicing in smaller and smaller increments. One second on this, two seconds on that. It's no wonder we take our evening news in sound bites and our advertisements in 10-second time slots. It lines up better with the small morsel sizes into which we carve our lives, all in the name of getting more done.

Time-slicing, taken to an extreme, can actually damage productivity because it takes time to switch from one thing to another. The more switching time you need, the less time you spend on something. You squander the time switching from activity to another, time you could be getting something done.

Multi-tasking is just one myth. Let's look at a few other popular "time management" notions.

Myth #2
Getting more done is just a matter of being better organized.

While being better organized can't hurt, there's much more to it. Productivity in people is not the same as productivity in industrial engineering, so organizing materials and the flow of work is not the magic solution it can be on, say, an assembly line. There, a practiced, clever watcher over the linear assembly line process spots an obvious waste of time and corrects it. Production speeds up three-fold, and he's a hero!

Productivity in knowledge work isn't like that. Beyond merely being better organized, managing many priorities also means you:

- Are creative
- Take initiative
- Are able to focus
- Know priorities

We're going to cover all of that in this guide.

Myth #3
If you want something to get done,
give it to someone who
isn't busy.

"Oh, don't give it to him—he has too much on his plate already!" Ever heard that?

Fact is, busy people are doers. If people are seeking you out ("Hey, Bob, wait up, I have something I need you to do!"), it's a compliment! It means you get things done!

Of course, there's a limit. After all, there are only so many hours in a day.

Still, if you want something to get done, do you look for someone who isn't busy? (Even if you could find one!) Of course not. You look for people who are known for making things happen!

Myth #4
"Never put off until tomorrow what you can do today."

If I could resurrect any figure in history to have lunch with, I think it might be Benjamin Franklin. Imagine the questions he could answer!

But if we got around to talking about some of his famous axioms, like the one about not putting things off until tomorrow, I'd have to say, "Times have changed, Mr. Franklin. Sometimes it's a good idea to put things off a day."

You should know, for example, when your most productive times are, when your creativity is at its peak, and arrange it so you do your hardest work then. That could be tomorrow. So put off until tomorrow, when you're at your peak, what you could do today, when you're not. Planning how you'll spend your time helps you spend it well, and may mean putting some tasks on a middle burner instead of a front burner.

And last but not least—remember what we said early on?

Myth #5
Time is something you can manage.

In case we haven't already dispelled this one, here's one more reminder: Whoever said "Time and tide wait for no man" is as right today as he was then. There is no such thing as "time management."

You *can't* manage time. Instead, you manage:

• Tasks

• Energy

• Communications

• Yourself

. . . within the time you have, which isn't stopping for you or anyone else.

■○■○■○■

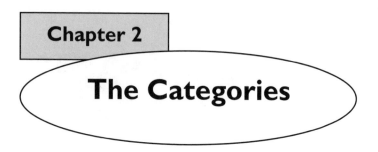

Chapter 2

The Categories

Managing yourself, your tasks and your energy means understanding all the ways in which you can better make sense of, and work through, the many priorities facing you. So we're going to approach this topic by talking about the several different categories that potential improvements fall into, categories that you personally can consider making some changes in.

If you're thinking you have no power to control what comes at you and how you get it done, you'll be happily surprised to learn that isn't true. Several of these categories are areas *within* your control. In fact, most of them are.

Of course that's not true for all of them. Sometimes you can't anticipate what's coming at you. I know. I live in the real world. I spent over 20 years working in Information Technology, in high visibility, hot-seat jobs, managing and sustaining mission critical systems. I know what it feels like to have a high priority system failure occur just as you're about to get in your car and meet your family at the Olive Garden for dinner.

So I've set aside a whole separate category for those things—"Beyond Your Control"—and we'll talk about what you can do (besides despair) to manage yourself and your tasks in those circumstances.

But for starters, we're going to talk about five categories that are about managing yourself, your time, creativity, energy and activities that *are* in your control and one (clearly named) that is not. Taken all together, the categories are:

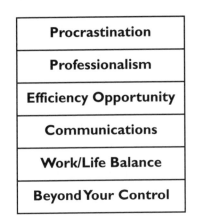

Procrastination
Professionalism
Efficiency Opportunity
Communications
Work/Life Balance
Beyond Your Control

- **Procrastination**—Why and how we put things off and how to get over doing so;

- **Professionalism**—Bringing your best self to work every day and how that helps you get things done;

- **Efficiency Opportunity**—The time drains that draw us away from real work; how to head them off at the pass;

- **Communications**—How to optimize the written and spoken word to get more done;

- **Work/Life Balance**—Work and life aren't locked in a tug of war; how to see your work and life in a continuum, not a deadly embrace;

- **Beyond Your Control**—How to influence and improve the things that just come at you to help manage your circumstances.

For each of these categories, we'll look at:

- **Overview**—What's included in the discussion of each category;

- **Symptoms**—What it looks like and feels like to have problems in this area;

- **Causes**—The reasons behind problems in each category;

- **Recommendations**—What you can do about them.

At the outset of this adventure, there's an assessment worksheet for you to complete. It's called **Ask Yourself** and it will help you identify things about yourself in each of the six categories that will let you better manage many priorities.

There is also a **Personal Plan,** with sections for each of the Categories we're examining. You can build your plan as you proceed through this guide.

We'll talk, too, about the **Productivity Pillagers.** These are, according to research in "time management," the four most widely-acknowledged "time management" problems. They are:

- *Interruptions*
- *Email Overload*
- *Ineffective Meetings*
- *Negative Thinking*

Notice that, of the four **Pillagers,** only one falls into a category that is "Beyond Your Control":

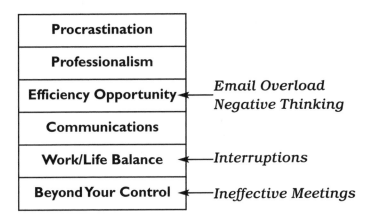

Procrastination
Professionalism
Efficiency Opportunity ←—— *Email Overload* / *Negative Thinking*
Communications
Work/Life Balance ←——*Interruptions*
Beyond Your Control ←——*Ineffective Meetings*

Where We Go From Here

Managing many priorities depends on knowing what Categories you can improve in. To do that, you'll complete an assessment—**Ask Yourself**—to pinpoint the areas you need help with. Then you'll read through the recommendations and adopt those best suited to you. These, you'll put in your **Personal Plan**.

Easy, no?

Ask Yourself

Step 1:

In the Appendix section, you'll find the **Ask Yourself** assessment. Without looking ahead (you're on the honor system here!), fill in the assessment, per the instructions on the top of the page.

Step 2:

After you've completed it, turn to the **Ask Yourself Score Sheet** and put your score in the designated rows for each section.

Step 3:

On the **Ask Yourself Score Sheet,** record any 1's and 2's you have. Now notice what Categories you gave yourself 1's and 2's in. Those are areas of particular importance—for *you*. They indicate opportunities in those areas to adopt the specific recommendations in this book to help *you* manage many priorities!

■○■○■○■

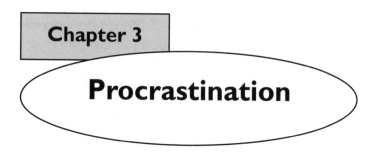

Chapter 3

Procrastination

We're going to start our examination with **Procrastination.** I mean, we shouldn't put that off, now should we?

Procrastination is one of the five categories *within your control* in which you, personally, can effect improvements that will help you get more done.

Overview

There is such a thing as a normal, healthy postponement of something you don't want to do. It would be crazy to expect everyone should dive right into every miserable task facing them, even when it's good for them.

You'd think twice about someone who said, "Gee, I can't wait to get my flu shot. I think I'll get up early in the morning and see if I can be first in line!"

That's not procrastination. Putting off unpleasant tasks is normal.

Procrastination means putting off reasonable things to do for so long that it interferes with your productivity, your ability to meet deadlines and your sense of feeling good about yourself and what you do.

It can be even more severe than that. Procrastination can be bad for your emotional health. If you're really a procrastinator, you may find you question your own motives. "If it's so important to me, why don't I do it? What's the matter with me?"

Procrastination can even damage your social life, if you find your last-minute approach to getting things done takes over your calendar and you're cancelling things you'd like to do in favor of things you have to do because they're overdue.

Procrastinators can find their credibility is damaged. "Don't give it to Rob to do, unless you want it thrown together at the last minute!"

That's what procrastination looks like in its more extreme forms. The truth is, if you're a procrastinator, you probably know it.

Who does procrastinate? According to Dr. Tim Pychyl, a Professor of Psychology at Carleton University in Ottawa who's done quite a bit of research on the subject of procrastination, 20% of adults and 70% of college students do.[1]

So it's a common problem, but maybe the fact that a much higher percentage of college students do so than adults is a good sign! Perhaps many of us grow out of it.

Symptoms

What does procrastinating look like?

- **Work Avoidance Behavior**—We've all done this at some time or other. That term paper was due, but instead of working on it, you were cleaning your apartment from top to bottom. Man, that place was clean when you were done! If only you'd written the paper—which was actually *due.*

  Or you go looking for something in your desk, the guidelines for the annual budget, so you can work on that long-postponed task of preparing your department budget. As you rummage through one pile and then another, you begin filing—something you haven't done for two months—and before long you've made quite a dent in the filing. But you haven't started on the budget assignment you're still avoiding.

- **Unrealistic Optimism**—Ever asked someone to estimate how long it will take to, say, complete testing on a product component and he or she said "a day, maybe two," but in your head you thought it would be more like two weeks? It may be a procrastinator who's giving you this unrealistic estimate. This is a way of not dealing with something we don't want to do: Assume it's easy and won't take long, so we can put off getting to it and think we can still get it done on time.

✓ **"I do my best work under pressure!"**— Sometimes we're such practiced procrastinators that it's simply how we do things. We think it's normal. We justify it. A practiced procrastinator says "I do my best work under pressure" because he has adopted the fallacy that, by putting things off, the quality will be higher because the deadline is imminent. He may, according to Dr. Pychyl, even enjoy the self-imposed challenge of the last minute frenzy.

 ## A Note About Working Under Pressure

"Hey wait! I really do my best work under pressure! I always have!" Is that what you're thinking?

Really? Think about it. Are you really saying that when you make sure you *don't* have time to make last-minute corrections that you produce a *better* product? Do you honestly mean that you're always at your optimal creativity when the clock is racing beside you? Even if it's the middle of the night and you know yourself to be a morning person? Not just that you were once under those circumstances, but that you are, all the time, equally inspired and productive because the countdown is on?

Last but not least, are you sure that when you don't have time to supplement your work with additional research, information, or get a quality check from someone else, that's really your very best work?

Hmm. Reconsidering? Maybe you do your *only* work under pressure. Or maybe you like the excitement you get when the clock is ticking and the pressure is on. The harm in letting yourself enjoy the rush, though, is that you'll create the crisis just to enjoy the rush and, in the process, not live up to your own expectations for yourself.

Causes

Why do we procrastinate? Certainly, to some degree, it's normal. We're only human, after all, with a tendency to put pleasant things ahead of unpleasant things, to get at doing things that come easily before doing things that are harder.

But beyond the "normal" reasons, what are some other causes?

✔ **We're Uncertain**—about the task. As a manager, I often used to put off getting at my monthly financial updates because I was never sure where to begin. I'm not an accountant, and I found the statements (and the accountants!) hard to decipher. Every company was different, with its own way of doing them, so job changes only made this worse. I'd wallow in uncertainty, often until it was on the verge of being overdue. I could always find something valuable to do instead that I did understand.

✔ **We "Lump"**—we don't know where to begin. Large assignments are especially vulnerable to this one. "This is *huge*," we think to ourselves, "so for now I'll work on something smaller, maybe even get that smaller thing off my 'to do' list and that'll free up more time to figure out the big guy—later!"

✔ **We Worry**—that "it won't be good enough." We believe we can do a good job, but not one worthy of the importance of the assignment. If we never start in, then the promise of perfection is still out there, Dr. Pychyl says. Once we get going, the imperfect job we're going to do is underway—just as we predicted: Not good enough.

Recommendations

Ask for help: If you recognize that feeling of uncertainty as you face a task, consider this: In work cultures, asking for help is complicated. We don't like to bother people—they're busy. We don't like to look like we don't know. Maybe the boss will take asking for help as a sign of incompetence. We're very practiced at going it alone.

But the fact is, when you do ask someone for help, usually they say "Sure." Think about it: If someone asks you for help, aren't you usually more flattered than bothered? What's true for you, in this case, is usually true for others. I even found that to be true with the accountants who had to help me untangle spreadsheets to make sense of my annual budget.

Keep in mind, too, that what's unclear to you may be unclear to others. They may be just as unwilling to bring it up as you are. So ask for help.

Just Start: You take a long trip one mile at a time. You eat Thanksgiving dinner one bite at a time. Approaching a huge assignment is the same thing. If you "lump," start with one bite, one step. Start by figuring out where it makes sense to start. Decompose a large job into smaller ones, and then start.

Try "time-boxing"—*that is, set aside half an hour which you will devote to this huge, daunting assignment. Use the 30 minutes exclusively to think about and work on whatever seems too big to begin. At the end, you may find you've broken down your own resistance and you want to keep going. But if you don't, and you're grateful the time is up, then stop. Make another 30 minute appointment with yourself for later to work on this some more.*

Re-set Your Expectations: Sometimes "perfection" *is* required. Ask yourself: Is this one of them? If it isn't, set your expectations accordingly and get on with it.

If you're prone to worrying, consider that not everything has to be done to the same level of detail, and many of us are in jobs where we iterate. I draft something, give it to you. You look it over, revise, give it back to me. If that's the case, use the iterative process to improve what you started, rather than think you have to make it flawless the first time.

It's important to make good decisions in this area. For example, I would want to think that a problem with some engineering design flaw on a commercial aircraft had been fixed, more or less, to perfection. I wouldn't want to think it had been compromised because the person working on it was trying to get over his procrastination problem.

But it isn't always the case that everything you do has to be perfect. Knowing when to let go a bit—and then doing so—is important.

 ## Using Emotional Intelligence to Conquer Procrastination

In the last decade or so, there's been a lot of talk about a phenomenon known as "emotional intelligence," the personal and social abilities of individuals to recognize emotions and use them in their actions, decision-making and interactions with others.

Emotional Intelligence (EI) is the hard science of soft skills, built on a physiological understanding of the brain chemistry behind emotional behavior. Although it's been around awhile, it was popularized by psychologist and consultant Daniel Goleman, whose book *Working With Emotional Intelligence* demonstrates how useful EI is on the job.

One area of personal competence within EI that Goleman talks about is the ability to "self-regulate," applying self control to ensure you keep impulses in check.[2] If your natural tendency is to work yourself into a froth trying to achieve perfection, recognize this is an opportunity to control the impulse, and then do so. Use your EI: Self-regulate!

If you're thinking to yourself, I've never been good at this sort of self-control thing before, take heart! Goleman says that, as we get older, our EI improves! Unlike IQ, which is relatively fixed for most of your life, EI grows as we age. Something about the physiology of the brain actually allows us to get better at these essential areas of competence that complement our IQ.

Finally, some good news for anyone who's worried about getting up there in years. Your eyes may fail, your back may give out, but your EI is on the rise!

One action item that can spring from this is that you make a pact with yourself to try time-boxing, try asking for help, just start, etc. If you're self-regulating, you'll keep that commitment. Who better to keep an important promise to than yourself?

One More Tip

Collaborate. Consider turning to someone you trust at work and tell him or her you're actively chipping away at your tendency to procrastinate. Say what you're doing, specifically. "I'm going to let go of the idea everything has to be perfect" or "I'm going to set aside 20 minutes to take a look at an assignment I don't want to do." Perhaps you can ask this person to point out when you seem to be backsliding on your commitment to yourself. It can help to reinforce your resolve.

Think about it: If someone turned to you for the same kind of help, would you think less of him or her? Would you say no?

Your Personal Plan

In the Appendix section in the back of this book, behind the **Ask Yourself** Assessment you did, you'll find a **Personal Plan,** a form for you to use as you work your way through this guide. Now that you've completed the "Procrastination" section, flip to the **Personal Plan** and jot down in the space provided whatever tips and techniques resonated with you in this section.

Consider answering these questions in your **Personal Plan:**

What improvements are you going to make?

Will you be pulling back from perfectionism?

Will you take "one bite" out of a huge assignment you're avoiding?

What commitments are you going to make to yourself?

Will you re-think your "I do my best work under pressure" approach?

Are there any work assignments facing you now that you'll try "time-boxing" on?

Whose help will you seek with work you're uncomfortable tackling?

■○■○■○■

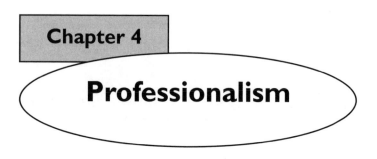

Chapter 4

Professionalism

Overview

When we're impressed with people on the job—colleagues, management, anyone—it's sometimes hard to put a finger on just why that is. It's their demeanor. It's a combination of their dependability and expertise. It's their character and it shows up, sometimes, in their status, or at least in how they're regarded by others.

See? It's hard to pin down exactly, but really what it amounts to is *professionalism.*

Wikipedia says professionalism is "The standing, practice, or methods of a professional, as distinguished from an amateur" and "professional status, methods, character, or standards."[1]

If it's difficult to characterize it, maybe we should think about what it looks like. People who have it:

- Are good at follow-through
- Remember to call you back when they say they will
- Are on time
- Do what they say they're going to do, keep commitments
- Are relatively organized

Emotional Intelligence and Professionalism

In Emotional Intelligence terms, this shows up as a personal competence: Conscientiousness, which Daniel Goleman defines as "being punctual, careful in doing work, self-disciplined, and scrupulous in attending to responsibilities."[2]

Expectations

So we have an idea now about what "professionalism" is all about. Most people demonstrate professionalism, some more than others. Only the most saintly among the professional workforce could demonstrate all aspects of professionalism all the time. And let's face it: If they were that perfect, we'd be suspicious.

With the right natural ingredients and the right influences, professionalism develops in people over time. Entry level people often demonstrate less than experienced professionals. That said, we've all seen examples of people who've been on the job a long time and have precious little.

The assumption here about professionalism isn't that everyone falls short and everyone, therefore, is due some amount of criticism. However, to discuss this topic we have to explore some personally negative things—undesirable behaviors and outcomes that show up in people. But let's do so in a framework that has as its underlying assumption that most people get up in the morning planning to do the best job they know how.

Yes, there are evil professionals, but they're rare. Most people don't drive to work every day thinking, "I sure am glad I promised that report would be done today because I have no intention of even starting it!"

Ask Yourself—Results

Most of us fall somewhere between incompetent and extraordinary and, depending on the job at hand, we'll be closer to one end of the continuum or the other. As you examine the results from the **Ask Yourself** section that fall into the "Professionalism" category, don't be hard on yourself. Keep in mind that most people come to work every day to do the best job they can. That includes you.

Dr. Richard Farson, in his wonderfully insightful book about paradox in business, *Management of the Absurd*, says it best:

"Most employees are trying to do the best they can. They prefer to do good work, to cooperate, to meet objectives. They prefer harmony over conflict, action over inaction, productivity over delays. Not everyone, and not all the time. But in general, people want to perform effectively."[3]

Now, if you've fully internalized that, turn to the **Ask Yourself** results (see Appendix—**Ask Yourself Score Sheet**) to see where your opportunities to improve your professionalism are.

Symptoms

How do problems with Professionalism show up:

- ✔ **Missed deadlines, late to meetings—** We all know people who are consistently late to meetings. We schedule lunch with them 15 minutes ahead of when we actually plan to start, so we don't get too hungry waiting for them. We may be less aware of missed deadlines on the part of our peers, unless we're working on something together, but news about missed deadlines often comes out—either as a confession or because we discover some upstream dependency that wasn't met.

- ✔ **Undependable; lacking follow-through**—There's a difference between missing a deadline and not meeting commitments at all, not coming up with the goods once assigned. Follow-through, living up to your accountability, is an essential ingredient of professionalism, and consistently undependable people aren't long for their jobs.

The all-promise-no-deliver form of this is probably the worst. "I'll do it!" earns the volunteer the immediate gratitude of the meeting leader, unless that person has a reputation for always stepping forward at first and then consistently not coming through. These habits waste everyone's time.

✔ **"I've got to get organized!"**—Staying organized is a personal thing. Anyone can make an edible meat loaf by following a recipe, but unfortunately there's nothing quite as precise as that for staying organized. Each well-organized individual has his own recipe. For some, it means the filing must be done every day. For others, 200 unopened emails waiting to be read doesn't affect their ability to deliver. For yet others, a neat and tidy desk is required. But being neat is not the same as being organized.

Neatness Counts?

I once worked for a man who was meticulous. He always placed chairs under his conference table equidistant apart, pushed in just so far.

He insisted on erasing every mark from the white board in his office after every meeting. On Fridays, he wiped down the same board with cleaning solu- 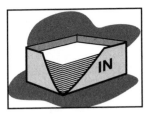 tion 'til it shone, removed any dubious-looking leaves from his office plants, and stacked the papers in his inbox so that every corner was aligned with precision.

Although he was fastidious, the technical staff who worked for him were not. One week, tired of the relative states of chaos in the cubicles of those who worked for him, this vice president declared Cubicle Clean-Up Week and promised a prize for whomever he deemed most improved.

Some people took him seriously, others didn't. One who didn't, particularly, was the department guru, a guy whose desk was often hard to find amid the mess he lived in. But, for fun, the guru re-cycled some of what he'd collected and then crammed the rest into drawers so his workspaces were cleared off.

That Friday, he won the prize.

On Monday, he unloaded his drawers back onto his desk, and most of the damage was restored.

The guru was known as an expert with great follow-through skills and a quiet, optimistic manner—a real professional. He never dropped an assignment, never missed a deadline, and the quality of his work was considered exceptional.

Was he neat? No, but he was organized.

A Note About Clean Desk Policies

Some companies have "clean desk" policies which usually mean that, at the end of the day, or certainly at the end of the week, the tops of everyone's desk should be clutter-free. Things should be in drawers, put away, maybe even locked up. Often, this policy is in place for security reasons, keeping confidential or proprietary information away from surfaces where they may easily be observed.

These days, in many places, the cubicle prairie is protected behind locked doors, so there's not quite the sense of urgency about clean desks. Nonetheless, a clean desk policy is not the same thing as a company-wide "get organized" mandate. Being organized in a professional setting is, as we said, highly personal. Attempts to enforce it usually fail.

Causes

If you've identified some areas where you, personally, are falling short in the category "professionalism," it may be attributable to one of these:

- ✔ **We're Inexperienced**—Anyone new to working in a professional setting has a lot to learn, everything from "how we do things around here" to what's acceptable on Friday afternoons and how best to communicate with the boss. That's just a fact of life. Add to that the need to build your expertise and polish up your demeanor, and that's a lot to get right. It takes time to develop professionalism.

- ✔ **We're Unprepared (or Underprepared)**—for the job we're asked to do. It seems like we're flaking out, but the truth is we don't know where to begin or what to do. Rather than admit we're not sure, we don't deliver. Sometimes, if we've done well in one area, we're asked to do something brand new in another area, success being its own curse, in this case. Consider that you may be underprepared for the assignment at hand *because you've done well.*

Under-prepared for a New Role

In a software development shop, a talented programmer was asked to participate in testing the completed system. She'd never been a tester, but it seemed like a logical area for her to jump right into and be good at. She agreed—until she saw the forms she had to fill out, which made little sense to her. Up until then, she'd been designing screens and retrieving data. Now she was being asked to think like a business person and determine what "expected results" this system should provide.

"How should I know?" she wondered. "I'm a programmer, not someone in the business who's going to actually *use* this system."

Her success as a programmer landed her a new assignment, for which she was unprepared, but no one knew that—until she tried to perform.

- ✔ **We're Overwhelmed!**—Failure to deliver—on time or ever!—may be the result of sheer overload. At a more manageable workload level, important elements of professionalism would be intact, but it's hard to keep your chin up and do your best when you're constantly swamped.

- ✔ **We Have Bad Habits**—Yes, difficult though it may be to face, sometimes our own patterns of behavior are so entrenched, they're bad habits. Like the person who always leaves his office for a 1:00 meeting at 1:00—so he's guaranteed to be late! Or the terrible typist who, out of habit, never proofreads anything. Guess what his finished product looks like when it gets where it's going! Not professional, that's for sure.

What's behind a lack of professionalism may just be weak personal standards.

Recommendations

Partner Up: Partner an inexperienced employee with an experienced one, perhaps in a mentoring relationship. Closer supervision (not micromanaging, just more assistance) will also help, but only when the supervisor sees the problem for what it is and focuses on the individual's need to learn about and adopt professional habits. Getting organized is a skill many who are new to the workforce don't have at the level they need for the job!

 ## A Note About Mentoring

Mentors can help other professionals grow their abilities in the important areas of follow-through, keeping commitments and being on time.

Although mentors have been around forever, a more formalized approach to mentoring has gained popularity in professional settings. In these mentoring programs, executives are asked to set aside time to spend with middle managers who, in turn, are asked to do the same for supervisors, prospective managers and individual contributors. Periodic meetings are set up, and some mentoring programs require that participants keep logs, journals or other documentation.

If doing all this helps to get people together, then there's certainly some value in that. But the most important thing to keep in mind about mentoring is the relationship, and that's impossible to predict or engineer. Setting up mentoring relationships is like setting up blind dates. It's not enough that the chosen mentor is an organized, efficient user of his own time and talents, or that he's a good teacher, or that he has other good examples of professionalism to share. What matters is that the two people in the relationship *like each other.*

Successful mentoring relationships can't be manufactured. They can only be encouraged. You wouldn't ask your blind date to go steady. Nor should you expect every mentoring relationship to blossom into something long-term. Despite the best efforts of matchmakers who arrange mentoring relationships, any mentoring program should set an expectation at the outset that it's okay if either person decides after a couple of meetings to part friends and try another mentor.

 **Develop New Skills:** If you're unprepared (or underprepared), the obvious answer is education. Someone has to explain it. Could be an in-

structor or simply someone experienced in whatever it is. The programmer who was asked to be a tester could have benefited from some hands-on experience in the business area in order to really understand what the system was supposed to do. That, combined with classes in, or reading about, what constitutes a good test case (like what "expected results" are supposed to mean), might have done the trick.

Speak Up: This is dangerous territory. Just because you're feeling overwhelmed, it may not be clear to your boss whether you're genuinely over-committed or just saying so. If you know you're operating at your best efficiency but still you're overwhelmed because you're permanently over-committed by forces beyond your control, talk to your manager. This may be a symptom of a problem he should be, or already is, fixing.

This is where it can get tricky. Cost-cutting measures have left a lot of companies coping with the promises of "doing more with less." Try to get an accurate read on how things are where you are—by asking your manager (not accusing)—and by not only seeking help with your workload but by also committing to making the improvements within yourself that will help not only you but the company.

Self-Help: If your professionalism could be improved if you weren't as overwhelmed, try some self-help—like the book you're reading right now! Take a class in managing priorities (or "time management"—if you must!), then be sure you make some changes, or you'll be right back where you started.

Using Emotional Intelligence to Reinforce Your Resolve

Fixing "bad habits" you've been carrying around for years requires applying Emotional Intelligence (EI). In particular, it calls on the EI competencies Self Control, Trustworthiness and Conscientiousness. Professionalism requires that when you make commitments, you keep them, and you do so because you hold yourself accountable.[4]

"I'll get it done by Friday" means you get it done by Friday. "I'll meet with that group to resolve our differences" is a promise on which you deliver. To be sure, this is not nearly as simple as flipping a switch. There's a lot more to it. The first ingredient is *thinking it through* before you commit. What will it take to get it done by Friday? How will I approach that other group, and do they want to resolve our differences or will they stonewall? If they stonewall, what will I do? (Don't just pose the questions to yourself, by the way. You also need to answer them.)

Work in a knowledge industry is about interacting with others. Meeting commitments means gauging how those interactions are going to go, how much you can do on your own, how your progress will be affected by others. Then you can be a person of your word, demonstrating your mastery of the EI competencies Trustworthiness and Conscientiousness.

A Note About Getting Organized: Task Yourself

Although being organized is a relative state, there are some essential tools and skills everyone should have in his trick bag. One is a task tracker. This is really nothing more than an enhanced "to do" list on which you keep a running list of what you have to do, for whom, when it's due, and how important it is.

It doesn't have to be a complicated tool. In fact, it shouldn't be. Its purpose is to remind you of what you have to do, by when, and to organize the list in order of priority. That's it. Any more than that, and you'll spend more time managing the task tracker than doing the work.

Task Yourself is just such a minimalist tracking tool, with columns for priority, task description, due date and miscellaneous notes. An example of the full tool appears in the Appendix, but essentially it's this:

Priority	Sequence	Project	Task	Due	Notes
a	1	Alpha	Test	1/6	Resume
b	1	Wichita	Fix line down	1/9	Call vendor (800) 511-1111
a	2	Work Tracking	Product evals	1/29	
c	1		Call Dr. for appt.	2/15	Ofc moved—need phone #
b	2		Write up self eval	1/15	
b	3		Teacher conf appt	1/20	Set appointment
a	3	Work Tracking	Prep for meeting	1/7	Need ppt slides from Alice

Priority describes what overall urgency you believe this task group has, whether this task is on the "front-burner" or "back-burner" (or, as an old boss of mine used to like to say, "fell off the stove altogether" for things that had been entirely forgotten!). In this case, the Alpha Project is a higher priority than the Wichita Project.

"Sequence" means, within the overall urgency, which needs to be tackled first, second, third, etc. "Project" can be listed, if there's a project associated with the task (might not be one if, for example, it was a personal task, like "buy birthday gift"). "Task" is the task description itself. "Due" is the date it must be done by and "notes" is a freeform area for short notes to yourself pertinent to the task you've listed.

If you thrive on complexity, you can certainly expand this simple tool beyond these few points. You could, for example, add a "status" column or a "date started" column, even a progress measurement. You could turn this single page into a project plan unto itself, if that appeals to you. The danger becomes that you spend more time in task tracker upkeep than in task execution. Personally, I recommend keeping it simple.

Using **Task Yourself** well means knowing what's important to you, to your job, your company and your life—and not necessarily in that order! Do you know what's pending? What's due? When it's due? Whom you need? How to reach them? Jot it down.

Planning and monitoring your tasks is the first step. The second, of course, is following your plan. The most ingenious planning and tracking system in the world is an utter failure if you don't *follow the plan.*

Another Note About Getting Organized: Clutter

If you wish you kept a neater workspace, try one simple technique: Avoid the "put-it-there-for-now" syndrome.

We're all guilty of it. "Well, I just put it here *for now* and then I'll put it away (handle it, file it, do something intentional with it) *later.*" Then later never comes.

Ever been in a teenager's bedroom? It's a living laboratory of "put-it-there-for-now" syndrome!

You won't always have time to properly dispose of, or handle, every item in your physical possession, but if you're aware of how often you put it somewhere "for now," you'll start to check yourself, and things will end up being put away the first time more often.

One Last Thought

If you are a seasoned professional for whom shortcomings in professionalism are less of an issue, consider offering to mentor those who have less experience.

Supervisors can set expectations that include demonstrating organizational skills. They should follow up by coaching and encouraging performance in this area.

Personal Plan Update

Now would be a good time to turn to your **Personal Plan** in the Appendix section of this book and write down any tips that resonated with you in this chapter that you want to remember and use in the future.

Consider answering these questions in your Personal Plan:

What improvements are you going to make?

Would you consider asking for a mentor? Or being one?

Are you going to adopt the **Task Yourself** *tool to help you track and manage the many things you're working on?*

Are you going to make a resolution to minimize your "put it there for now" approach to keeping your space neat?

■○■○■○■

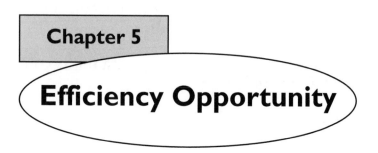

Overview

How well do you know yourself? Can you answer these questions?

- When are you most creative?
- How do you boost your own creativity?
- When does your attention span expire?
- Do you sweat the details? Should you?
- What doesn't stress you out (that does stress out others)?
- What do you know you spend too much time on, or not enough?

All of these represent opportunities to improve your personal efficiency.

Simply put, Efficiency Opportunities are:

- Recognizing the ways you *don't* optimize your time and energy . . .

- Knowing the best ways you personally can and should optimize time and energy . . .

- . . . and doing so!

Oh. Is that all?

Symptoms

Opportunities to improve efficiency look like this:

↙ **"I spend hours every day reading email"**—Remember the **Productivity Pillagers** we talked about in the Introduction?

The **Pillagers** are, according to research in "time management," the four most generally-acknowledged workplace problems people have. Being swamped by email is one!

↙ **Focus Problems**—Some people respond to competing demands for their time and attention by letting the frenzy run them over. Then they find they can't engage quickly in what they need to do, or can't disengage from what they've been doing so they can get on to the next thing.

✔ **It takes longer than it should to get things done**—You've been trying to figure out what's causing a particular problem for hours, and you can't. It doesn't seem like it should be that difficult. Or you've read the same paragraph in a document three times. Or you've been working on your status report for two hours when it should take you 30 minutes.

✔ **Negative Thinking.** This is another of the Productivity Pillagers!

Being discouraged from time to time is natural. But spending a lot of time being discouraged and expecting the worst is not, and it drains energy that could otherwise be spent productively doing work.

Negative thinking also shows itself as being critical of other people or other groups. "I never know what they're doing *over there,* but it sure doesn't look good!"

Causes

Missed opportunities for improving efficiency can be attributed to only one thing, I'm afraid:

Human nature! Because we're only human, alas, we're:

- ✔ **Not optimizing our own talents well,** not thinking through whether this is the best time to problem-solve, communicate, organize, read, etc. Instead, we simply hop to it, taking on more and more, pretending we're handling it by multi-tasking—which, as we now know, is a myth and is at odds with focus and productivity.

- ✔ **Seeing everything as URGENT and important** even when it isn't! That certainly applies to email. How many "urgent" emails do you get in a day? Why is that? Because everyone wants top billing, high priority, and email is immediate. It's also easy, cheap and convenient. It's not hard to get on to email, it's easy to send one, and it's possible to get instant action. No wonder it's so compelling! Its convenience and price have brought it into our midst with a vengeance, as anyone who slogs through his spam knows!

  Beyond email, there's even more confusion in the ranks about URGENCY. Of course it's true that some things *are* urgent and sometimes we must save the day! But when resources are scarce, one way to get something done is to declare it "urgent" even when it isn't, really, in the true sense of the word.

- ✔ **We're accustomed to criticizing others**—sad to say—because it comes more naturally to us than praising others or genuinely appreciating them, both of which make us squeamish. Culturally, talking down about others is an accepted "norm" in the workplace.

Negative Thinking

The Starbucks in my neighborhood is in and among a cluster of professional office buildings. I was in there one morning getting coffee when I overheard the conversation between two women in line in front of me.

"If you need something to get done, don't give it to Joan. Every time I give her something, I just have to do it over. It's not even worth it to ask."

"Oh, I know what you mean," the other woman said. "She's always been like that. Smiles a lot, but does a terrible job!"

As I stood waiting for my coffee drink, I overheard another conversation between two men at a nearby table.

"That whole department is weak. I don't know why they don't just clean house over there, get rid of the entire bunch of losers."

Wow, I thought to myself. Lots of negativity around here this morning!

As I sprinkled some nutmeg on my latte, I overheard another conversation, this time a group of three.

"We may not like having to work with them, but there's no changing it," the man in the group said.

"But they're soooo slooooow! We've been trying to teach them to use this software for weeks now, and they're just not getting it," a woman complained to him.

"I mean, geez," said the other woman in the group, "This is hardly rocket science. What's up with them?"

After that, I began to cruise around the crowded Starbucks, listening in on other conversations, and I heard one negative conversation after another, always about someone back at the office who was never any good at what they were supposed to do. I was disappointed but also enlightened: We encourage each other to talk down the other guy perhaps as a way of bolstering our own confidence. It certainly seems to be habitual. Everyone in my Starbucks experience fell into it easily.

But we pay a price for doing so: We squander our energy. Negative thinking is a waste of time.

Recommendations

Take Advantage of Your Best Hours: Rather than doing anything whenever or as soon as it hits you, try to take advantage of your most creative and energetic hours of the day. If you're not optimizing your talents well, consider rearranging your schedule a bit so you do your hardest work when you're "on," when your creativity is at its peak.

Are you a night owl or a morning person? Jeff Davidson in his book *Complete Idiot's Guide to Getting Things Done* is one of many researchers and writers who says a normal cycle looks like this[1]:

10 a.m.—noon: You're at your maximum mental skills
Noon: The dip in energy and effectiveness begins
3:00 p.m.: Alertness returns

Start working on a difficult, long or otherwise intimidating assignment when you're at your best, not when you're dog-tired and have nothing left to give. If you're a morning person, use the morning hours to your advantage. If a morning person waits until the sun is well on its way to the horizon, that intimidating assignment will take a lot longer to get done.

Morning People

At one company where I taught a class in Managing Many Priorities, I found I had a room full of "morning people." They complained that, every morning at 9:00, most of them had to attend a product status meeting. The information at the meeting was relevant, but they were pretty much passive observers.

"I can get so much done at 9:00 in the morning," one of them said, "but by the time I get out an hour later, I'm past that peak." Many people in the room agreed with him.

At a de-brief after the class, we talked about what people had said about the timing of this one meeting, and the manager who led the meeting agreed to move it to a different time. She agreed that optimizing people's energy was important!

■O■O■O■

More about optimizing your talents: When you have to perform some heads-down work that requires imagination (solving a problem, adjusting financial projections, creating a project deliverable, writing documentation), there are some ways to enhance your talents. For example, boost your creativity when you're "off" by asking for input, including brainstorming with others. Read or research what others are doing by looking in your own company-internal documentation or by going online. It's amazing how collaboration can re-start a sagging energy cycle.

Take a break—*You can't boost production simply by working longer and harder because knowledge work isn't like industrial engineering. If you need to come up with an idea or a fresh approach, or you simply need to do something you've done a hundred times but a little differently this time, just the inspiration you need to get it done might come from taking a walk around the block. Next time try that, instead of staring at it longer and harder in the hopes of revving up production.*

Separate Flame-Fanning from Real Fire Drills: If you have a tendency to *see everything as URGENT and important*, even when it isn't, take a moment to look for clarity about what's really an emergency and what's not. In the "do more with less" era that most companies find themselves in, getting things done involves a certain amount of arm-waving and jumping up and down: "Hey, over here, this is important! I need people, money, time!" Making priority calls gets harder to do, not easier, when resources are scarce—harder to do, but more necessary. Hence, the problem.

If it's not up to you to make a priority call, point out to whoever it is up to that there's a difference between on fire and important, and "Which one is this exactly?"

When it comes to responding to matters of urgency, another thing you personally can do to handle your email efficiently.

Managing Email

Long ago, early in my career, I worked for the University of Southern California in Los Angeles, leading a group that was supporting Personal Computers. In those days, our on-campus group sold PC's, serviced them and taught campus users how to use them.

Universities were early adopters of email, and USC was no exception. One regular user of email, who was also a PC owner, was the university's president, Dr. James Zumberge. He'd called us to say he was having trouble with his email when he dialed up (yes, it was in the days of dial-up) from home, and could we have someone come out to check out the trouble. My boss suggested I go.

"Me?" I said. "I'm hardly the team's top technician."

"No, but I think a manager should attend to the president," he said. Unconvinced, I nonetheless went to Dr. Zumberge's house to see what the trouble was.

I drove out to the estate in Pasadena, owned by the school, where the president resided. I was admitted at the gates, and drove up to the front door of this expansive, Spanish-style mansion. In the distance, dogs were barking as the front door swung open. Mrs. Zumberge showed me upstairs to Dr. Z's study, then brought me some tea.

There, next to his large stately desk, sat his ailing PC. I fired it up, and began an initial examination of the patient. Then I set to work swapping out cables and connectors, hoping that one of them was at fault.

While I was effecting repairs, Dr. Z himself entered the room. A tall slender man who'd been a geologist before becoming an administrator, he had the sort of serious, focused demeanor of a scientist. He asked if there was anything he could do to help.

"I do have one question," I said. "When you're in your email, and you're re-reading an email you've looked at previously, do you have the same problem both times?"

He tilted his head a bit to one side, as if the question puzzled him.

"I never open the same email twice," he said. "I read it once and move on!"

Chagrined, I said, "Oh, of course!" and thought, That must be why he's the university president, and I'm crawling around on the floor of his office swapping cables!

No matter what our job these days, email swamps us, so reading the same email over and over without responding, forwarding, deleting or taking some other action is a good way to squander time. The university president probably had an advantage many of us don't: He could delegate to others. Wouldn't we all like to forward an email to a subordinate with our own note attached, "Here, handle this"?

Even if you can't delegate (which, by the way, doesn't mean you're off the hook—you still have to follow up to make sure it gets done, even if you're the prez), I'm sure there are many times you've handled an email needlessly more than once. Try not to do that. Whenever possible, read it, decide, and move on.

Filter your email based on Subject. Use the "preview" function to see if this is really something hot, something for you, or just an "FYI."

Make your own emails actionable for others. Use "To" for recipients who must take action and "cc" for FYI or to indicate a courtesy copy, and say in your email that you're doing so. "If you're on the 'To' line in this email, please respond by Friday with revisions or corrections."

Because email is immediate, it's also compelling. I hesitate to say "addictive," because that really means something more, but some people do say it's addictive. For this, there's really only one remedy: Resist the urge to be in constant communication via email. Remind yourself that it won't hurt anything if you check in on your email two or three times a day instead of two or three times an hour.

If your job is to respond to urgent problems, and email is one way to be notified of them,

I'm betting that there are other ways—that you're paged, called on your cell phone or your land line. Trouble usually comes looking for you; you don't have to go looking for it.

Once you set a new threshold for email—like setting aside a few times a day to check it—you'll find all kinds of productive things to do with the time you're not shifting gears to dive into your latest email, some of which can certainly wait and some of which you can probably trash.

Punch up the positive—If you find yourself surrounded by Negative thinking and criticizing others, first you have to spot it, either in yourself or others. Try cruising around a Starbucks near business offices during a workday and see if you hear what I heard. Or simply listen to conversations during meetings, in the cafeteria where you work, or in the break room. If you hear negative comments, try inserting a positive one.

Compartmentalizing

For some people, the problem with negative thinking is that this morning's disappointment stays with them all day. They can feel the effects on their own productivity of being down.

If this sounds like you, try consciously *compartmentalizing* your outlook.

Ever been to a pediatrician's office? Except for the regular health check-ups, it's mostly one sick child after another, lodged in one small examining room after another. The doctor goes from one to the next, after stopping off to wash

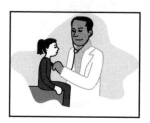

hands. It's amazing the doctor can (and, in my experience, does) devote his full attention to whatever child he's in the room with. He fully listens, he asks questions, he devises a plan. He writes a prescription, a lab order, some instructions. He's *all there* when he's there, even if he can hear the crying in the next room, even if he left an obnoxious child in the previous room. Whatever room he's in, he's all there.

That's compartmentalizing!

The doctor can move on because he has to. There's another child waiting and another anxious parent. You can, too—maybe not every time with equal composure, but certainly sometimes. The more often you do it, the better at it you'll become.

Emotional Intelligence and Compartmentalizing

You probably already noticed this, but the ability to compartmentalize is another way of applying Emotional Intelligence to the challenge. This is all about "self-regulating," an ability that shows up in part as an ability to manage "distressing emotions well" and to "think clearly and stay focused under pressure."[2] Many of us have these abilities to varying degrees and, thanks to our brain chemistry, this is an area we can build on and improve as we get older!

 ## Exercise: What's on your plate?

Try this. Take a few minutes to think about your strengths and preferences in terms of the areas we've discussed so far: Avoiding procrastination, professionalism and optimizing your efficiency. To help organize your thoughts, answer these questions:

Do you usually leave yourself plenty of time to get a job done, or do you like to save it up 'til the last minute?	☐ Plenty of time ☐ Last minute
Are you a planner or are you more spontaneous in your approach to work?	☐ Planner ☐ Spontaneous
What time of day is best for you? Are you a morning person? An afternoon person?	☐ I'm a morning person ☐ I'm an afternoon person ☐ I'm a night owl
Are you more naturally restless or naturally relaxed?	☐ Restless ☐ Relaxed
Are you meticulous about details?	☐ Yes, usually ☐ No, rarely
Are you comfortable on the phone talking with people?	☐ Yes ☐ No
Do you dread "difficult" phone calls or conversations, e.g., with unhappy customers?	☐ Yes ☐ Not my favorite, but I'm okay with it
Are you a writer? Do you like to write meeting minutes, technical documentation, how-to instructions? Or is that a "back-burner" item you hope someone else gets to do?	☐ I'm a writer, no prob'. ☐ I hate writing, it's a back-burner item for me!

Now imagine that the following list is what you have due this week. Think through how to optimize your energy and talents to accomplish the items on the list. What sequence could you do them in, which should you do early in the day, late in the day? Don't worry so much about the priorities of the tasks. Think about how you would optimize yourself to get them done.

CALL VENDOR about connectivity problems

MAKE A CLIENT HAPPY! Call customer to report successful resolution of problem.

MAKE A CLIENT ANGRY. Call customer to report that their problem is not yet resolved—still trying to figure out the cause.

UPDATE PRODUCT DOCUMENTATION to include new functionality recently implemented.

WRITE UP EVALUATION of new software products for Alpha project. Complete eval sheet, also short text describing answers.

WRITE STATUS REPORT for last week.

SET UP MEETING WITH SUPERVISOR to complain about how other team members aren't doing enough and I have too much.

PREPARE PROJECT UPDATE REPORT to present at our weekly staff meeting.

PERFORM PRODUCT TESTING—Execute test cases 95–109 and log results online.

WRITE SELF EVALUATION for annual review.

SET AGENDA FOR PICNIC COMMITTEE— Schedule meetings, recruit volunteers.

STUDY NEW SWITCH INTERFACE in vendor product documentation.

READ INDUSTRY JOURNAL ARTICLE the boss recommended in *Forward Thinking* magazine.

HELP NEW EMPLOYEE get up to speed.

If you're mentally at your best first thing in the morning, maybe that's the time you read complex technical material, especially if it's something you really need to learn and know. If negotiating with a vendor takes a lot out of you, do it when your tanks are full, not when you're already drained. If verbal negotiations come easily to you, and writing comes hard, get at those product evals when you have the most energy and tackle negotiations later.

Now, of course we don't live in this Utopian paradise where we can arrange everything we have to do in a sequence entirely to our liking. Obviously, there are forces that dictate when things are due that may be at odds with optimizing your energy and creativity. Nonetheless, within the constraints of externally imposed schedules and priorities, you can still be conscious of how to exploit your strengths and offset your weaknesses, just by optimizing your creativity and energy.

If you organize the tasks on the list, remember there's no one right answer here. The sequence should vary considerably from person to person because it should reflect individual talents, abilities and preferences.

One Last Thing

Know how you spend your time. You know the old saying: "If you can't measure it, you can't manage it." There's something to be said for that here. Spend a week tracking how you spend your time. De-brief with your own calendar every evening before you head home. Keep a running log on how you spent the last 24 hours, what percentage of time went to specific projects, to writing reports, commuting to and from work, socializing, eating meals, wasting time in meetings, participating in productive meetings—however you want to know where your time went. Do it for one week, and see how well your actual time allocation lines up with how you thought you were spending your time!

Personal Plan Update

Now's the time to turn to your **Personal Plan** in the Appendix section of this book and write specific tips and improvements that you want to remember and use in the future.

Consider answering these questions:

What changes are you going to make in terms of handling email?

Do you recognize elements of negative thinking in yourself that you'd like to re-direct? What changes are you going to make?

What changes are you going to make to how you prioritize the work assignments you have facing you right now?

Is there anything you'd like to talk with your manager/supervisor about that would help optimize your most creative or productive time?

■○■○■○■

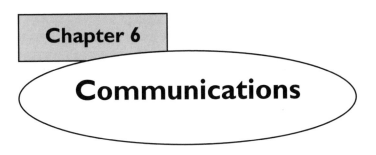

Chapter 6

Communications

Overview

These days, we hear a lot about "communication" at work. We hear how important it is to be good at it, how it's an essential job skill. How many job postings have you looked at that said at the end: "Excellent written and verbal communication skills required"?

We also hear, all too often, how terrible everyone seems to be at doing it. Seems like we have to go all the way back to President Reagan to find a "great communicator."

Engineers, scientists and healthcare professionals are supposed to be among the worst. Many people say they're terrible communicators, that they hate to write, they dread public speaking, they hate writing emails. Communication is regarded as an afterthought, a necessary evil. It's the cod liver oil, not the chocolate truffle.

Unfortunately, communication makes us think of documentation, conjuring up images of lonely, dull work, of typing and revising and wondering about word choice and paragraphing and grammatical correctness.

Communication also makes us think of interpersonal skills, especially our own, how well we do or don't look someone in the eye and say clearly what we mean, how well we say it in email.

Communication makes us think, too, of presentations, of a speaker in front of a room full of people, trying to conquer stage fright while PowerPoint slides fill a darkened room. These are some of the popular forms that communication takes in today's workplace.

But before we get too swept up in those conventional notions, let's talk about what communications are supposed to accomplish. Not what they *are*, but what they *do*.

No matter the form of communication (written, verbal, public, private, etc.), the goal of communicating is to:

- Get help
- Transfer knowledge
- Persuade
- Build consensus
- Calm fears . . . or stir up action

In a nutshell, communications are supposed *to keep things moving!*

So, then, effective communication is about:

* Making progress
* Minimizing re-work
* Sustaining the kinds of relationships that get things done!

Those are the ways that effective communications save time and help us *manage many priorities!*

Symptoms

It's easy to spot ineffective communications. They look like this:

✔ **Important material goes unread**—Ever have that feeling when you're writing something up that "no one is going to read this anyway"? You may be guilty of it yourself, knowing that something important is waiting in your in-box but you can't get to it now, or you don't want to. When there are fires to douse, who wants to read the boring material in the queue? It doesn't seem important.

✔ **Important info goes unreported**— Saying "All is well!" when it's not most often reflects a cultural issue. Maybe we think problems will reflect badly on us, so we postpone bringing them up as long as possible. Or we don't bother to write them down for the reason mentioned above—"no one will read it."

✔ **Infrequent, incomplete communication with manager or supervisor**—is a sure sign communications are ineffective. Regular communications, especially live and in person or, when necessary, via email, are essential to getting things done. Remembering that communication is a two-way street, the fault, in this case, could certainly be on both ends.

✔ **Unwelcome interruptions** (another **Productivity Pillager**) are a way communications are rendered ineffective, not because we can't respond well when we're interrupted (of course, some do that better than others), but because we're not using direct, diplomatic communications to postpone a conversation that's suddenly leaped onto our path.

We've all had them happen to us: Cube drop-in while you're in the middle of something; phone rings at the least opportune time; hallway moment—"Hey, glad I ran into you!"— while you're headed somewhere; after a meeting—"Do you have a minute?" or in the middle of lunch—"Sorry to catch you midburger, but . . ." Not to mention the parking lot

after work or even in the restroom. If un-welcome interruptions are a regular time drain for you, you need to adopt some tactful, straight-forward communi-cations techniques to intercept them.

Causes

✔ **"Writing (or speaking) is not my thing."** A couple of decades ago, educa-tors in schools often said "Girls aren't good in math." While we've stopped say-ing that about girls and math, it doesn't seem that anyone has realized yet that we've instead shrouded communications in the same veil of difficulty, so much so that "I can't write!" is a phrase too many people believe about themselves!

✔ **Communications are boring, and it isn't "real" work**—Real work is respond-ing to emergencies, competing in the global economy, building, fixing, testing, discovering—not writing it down! At least that's the conventional thinking. To com-pound the problem, many organizations have created templates—some of them very rigid—for standard communications.

Some standardization is good, but it is possible to have too much of a good thing. When communications are overly "templa-tized," it can wring the life out of a mes-sage. Readers stop reading, glaze over and move on to something more fun to do.

🖋 **Complaining is frowned on**—Coming forward with bad news is considered by some to be complaining, and many people think they know better. Why bother? What good will complaining do if my boss's attitude is "Don't ask for help—it's up to you"?

🖋 **"I hate to say 'no.'"** Why do we let people interrupt? Because they have an urgent need. Or because we like them, or maybe we're too polite and we think "no" is a form of rejection. Maybe being interrupted makes us feel important. Or maybe we let people interrupt because we just can't think of what to say to postpone the moment to a better time.

 ## Communications and Emotional Intelligence

In Emotional Intelligence terms, communication is, not surprisingly, a social skill (as opposed to a personal skill). It's about "effective give-and-take," and dealing "with difficult issues straightforwardly," staying "receptive to bad news as well as good."

The ROI on Communications

Before we discuss specific recommendations you can adopt so that better communications practices help you with your conflicting priorities, let's make sure we're in agreement about one thing: There's a hard dollar Return on Investment (ROI) on effective communications because they pay back in time savings. If you eliminated all the occasions you had to clear up confusion because someone didn't get it, or if you never again had to say the same thing over and over because your explanation fell short the first time—wouldn't that save you time? If you could hand someone a completed how-to document for something that only you knew how to do, wouldn't that cut down on the amount of time you personally would have to spend explaining it?

You get the idea.

Moment of Insight

In my career, I've interviewed dozens of people, maybe hundreds. But the best answer I ever heard in response to an interview question came from a college senior who had applied for a position as an intern in our Information Technology department.

I was part of a panel of interviewers, and we'd put together a list of questions to ask all the prospective interns. One of the questions was, "When you're in a situation at work where you disagree with someone, maybe the disagreement even becomes heated, what do you do to resolve it?"

Most of the candidates said things like, "I re-state my point, making it very clear what my meaning is," or "I choose my words carefully so there's no misunderstanding what I'm try-ing to express."

But one young woman thought for a moment, and then said simply, "I listen."

We were stunned by the maturity and insight-fulness of her response. She went on to say how resolution would never be possible with-out an understanding of where the other per-son was coming from, and the only way to get there was to listen.

She had already learned the most valuable lesson of effective communications: It's not what you say, it's what you hear.

Recommendations

Write Because You Care: If you're thinking, ***"writing (or speaking) is not my thing"*** it may be because that you're feeling this way because of high school or  college experiences where you crawled through English 101, facing with trepidation the dreaded essay or, worse yet, the corrected essay, covered in red ink with marks impossible to internalize.

Not to mention that memorable Communications class, where you spoke in front of the whole class and now can't remember a thing you said, only that you were glad to sit down afterwards.

If experiences like that have led you to believe you have to be a polished presenter or that you have to know how to fix all those red marks in order to be a good writer, that's unfortunate—but not fatal.

In order to be a good communicator, the most important element is this: You have to *care* about what you're presenting. Expressing yourself well depends on whether it really matters to you that your audience gets it. This applies equally if it's in writing or in person. That's the key.

Don't get lost in thinking you have to be a qualified grammarian or an outstanding essay writer in order to be any good at communications. While there's no silver bullet solution, here are a couple of simple things—a different approach from the conventional one that took the wind out of your sails. These are things you can do every time to optimize your message and minimize the re-work time that poor communications always require.

Essentials of Communications

These are the three communications essentials that will help you optimize your message and minimize your re-work time.

Essential #1: *Consider Your Audience*

Like our young intern candidate (who, by the way, got the job), don't think about yourself. Think about *them,* your audience, i.e., whoever is going to read what you're writing or listen to what you're going to say. Think about what *they* need to know. What do they already know? What's in it for them to listen to you? It's your job to make the best use of their time.

Do they need to know something you know in order to do their job? To make a decision? To approve (or deny) money? Check progress? How much detail do they require in order to decide, approve, continue on?

That's where you focus what you have to say: On making it comprehensible and useful to *them.*

Essential #2: *Know Your Goal*

Begin your communication with the end in mind. What are you trying to accomplish? Answer this one question: How will you know, when you're done writing, speaking, talking with someone, that you've been successful? What do you want to have happen as a result? If you can't answer that one question, you're not ready to proceed.

The answer should be something like *I want people to:*

- *Take action!*
- *Understand something new!*
- *Be inspired!*
- *Start worrying!*
- *Stop worrying!*
- *Help me out!*
- *Work faster!*
- *Figure this out!*

Whatever it is you want, have it—and keep it—in mind.

Also, make sure you never leave your audience thinking "I'm not sure what you want us to do next." This is a time-waster you definitely want to avoid. If you want someone to read and revise, clear a roadblock, approve a budget increase or whatever it is—somewhere within your communication, and certainly at the end, say so!

Essential #3: *Anticipate Reactions*

Put yourself in your audience's place: How would you react to what you're communicating if you were them? Predict the questions they might ask. This is a huge time-saver because it keeps them from having to come back to you later with questions and concerns.

Be sure to anticipate more than just the factual questions, but include also the emotional reactions you think you might elicit. Are you delivering unwelcome news? Is what you're saying controversial? Are you likely to be seen as a trustworthy, qualified communicator on this topic?

Clear the air! Don't let doubt or unasked questions linger because they'll always surface later—when they take longer to answer or resolve. When you're optimizing your efficiency in this area, you don't want to be circling back with your audience afterwards to address any negative reactions that are in the way. The more you can do that up front, the faster and more directly your message will penetrate and the faster you'll achieve the goal you set.

Perk Up Your Work: If you're thinking, ***communications are boring, and it isn't "real" work:*** *Au contraire.* If it weren't for communicating, "real" work would never get done. Build a little time into planning your communication—using the three essential elements listed above. It will pay you back in time saved, guaranteed. If others still believe communicating isn't "real" work, too bad for them. At least you will know better.

To freshen up boring communications, here are a few ideas:

1. Avoid boilerplate language. Have you ever read a document where the first few paragraphs were exactly the same as something you read before? Project documentation is especially guilty of this, always starting every deliverable with precisely the same standard, approved sentences to describe the project that every other deliverable uses.

   "The Alpha project is a comprehensive re-design of our customer service system to enable faster call handling and improved data integrity. Using products compatible with our reference architecture, the Alpha Project will implement a leading edge solution that will . . ." and blah, blah, blah.

   As you sit down to read the latest work product from this project, you read those sentences for the twentieth or thirtieth time. Word for word, it's the opener everyone uses on every deliverable. You think to yourself, "I know that already!" Or you skip over it. After all, you've read that same description over and over, you don't need to read it again.

Why write something no one reads? Why start a document with an approach that's such a turn-off?

Say it differently, or don't say it at all. The standard project description is probably enshrined in some approved repository. No need to trot out the same old tired phrases, just for the record.

2. Make it lighter.

If you suspect people aren't reading important materials you really want them to read, try burying a sentence in there somewhere to get their attention: "First person to read this sentence, call Susan for a free latte at Starbucks!" You'll be surprised what competition is inspired by the prospect of winning a $3.00 coffee drink!

Shocking Feature!

A written test plan for a new insurance billing system was modified to include specific tests for a special feature. When a user lingered too long on a particular page, the system would send a substantial electric shock through the keyboard into the fingertips of the user!

This was, of course, a joke, and the test plan writer put it in just to cheer up his teammates.

But several revisions later, the joke hadn't been removed. The Project Manager, who had enjoyed reading about the special electric shock feature earlier on, didn't notice the bogus test cases were still in the plan when she circulated it to the stakeholders for final review and approval.

The test plan owner alerted the Project Manager that the version she'd sent out still had the electronic shock test cases in it. But the stakeholders read and approved it without commenting on the special feature. At first the PM was embarrassed, then she became concerned. Had no one noticed? She called around to the approvers and, without mentioning the specific tests, asked in several different ways whether they were okay with the final test plan.

After each of them assured her they were, she pointed them to the bogus test cases.

They became more attentive readers after that, never knowing when they'd be entertained like that again.

3. Update!

Stale news is especially a problem for status communications, so never say the same things over and over week after week, even if they're still true. Park old issues somewhere else, not front-and-center in a weekly status document where everyone is tired of seeing no progress (even if progress is stalled for legitimate reasons!).

Make Your Case Objectively: If you're worried because **complaining is frowned on,** be sure to state your case objectively, and with tact. Whatever you do, don't whine.

Sometimes people feel they have to punch up the negative in order to be heard. Often this happens when standard reporting (status reports, for example) have a reputation for being so boring no one reads them.

That leaves writers feeling they have no choice other than to make an overly long, impassioned case for something—to right a wrong, expose a risk, reveal an important but controversial truth. So, where complaints are realistic, this approach gives the whole thing a bad name.

Don't say "no" if you hate to say "no": Managing your way through this Productivity Pillager simply means saying "not now," in one of the many polite, respectful ways there are to say it. Here are a few:

"Do you have a minute?"	Say, nicely and with a smile, "I don't right now. Can you call me?"
	Ask "Is it urgent?"
	Say, "I need to be somewhere else and I'm late."
Cube drop-in	Stand up. (People will usually back away, start to leave.)
	Say, "I'm deep into something. Can we talk later?"
Phone rings	See who it is. Don't answer it. Note to self: Call back—or not, depending.
	If you answer it and you wish you hadn't, say "I shouldn't have answered this right now because I'm tied up with something. Can I call you back?"
Parking lot	Keep walking while you say, "I'm too beat to think about this now."
	"If this isn't urgent, can we talk about it later this week?"
Restroom	Dry your hands and go! Unwritten rule: No one is obliged to do business in the bathroom.

Think about it: If someone said any of these to you, wouldn't you say, "Sure, no problem"?— unless, of course, it were a genuine emergency, in which case you'd say so and an interruption would be warranted.

If you'd react that way, so will whomever you're talking to. Try it.

Personal Plan Update

Now's the time, once again, to turn to your **Personal Plan** in the Appendix section of this book and write specific tips and improvements that you want to remember and use in the future.

Consider answering these questions in your Personal Plan:

Who are "regulars" in my audience, and what should I keep in mind about them?

Do I allow myself to be interrupted too often? What practices will I adopt to help minimize unwanted interruptions?

Am I a good listener? What could I do to be a better one?

Am I aware of what goals I have for written communications?

For the things I produce regularly (like status reports, project deliverables), will I take a few minutes to focus them on an objective?

■○■○■○■

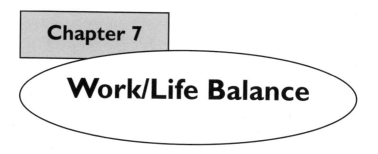

Chapter 7

Work/Life Balance

Overview

An Empty Pitcher

I've heard this story a number of times, so often I'm sure the original source of it has been lost in the frequent re-telling. My favorite version is that a commencement speaker is addressing a group of about-to-be Business School graduates at an east coast Ivy League university. As he looks out at their bright eager faces, he reaches under the podium and pulls out a clear glass pitcher.

"This empty pitcher," he says, "represents your life."

He reaches under the podium again and brings out several fist-sized rocks and puts them into the pitcher until the pitcher can hold no more. Then he shows the pitcher with the rocks in it to the graduating class.

"What do you think? Is the pitcher full?" he asks them.

"Yes!" they call back.

So he reaches under the podium again and pulls out a box filled with gravel. He pours the gravel into the pitcher. The pieces fall into position around the rocks, and he stops pouring once they reach the top.

"Now," he asks them again, "Is the pitcher full?"

"Oh yes!" They nod this time.

So he reaches under the podium and brings out a box of sand. He pours the sand around the fist-sized rocks and the gravel until it reaches the brim.

With a smile, he asks again, "Now is the pitcher full?"

The students look at each other, then at him. One young man finally says, "No?"

The commencement speaker nods and reaches back under the podium. He brings out a bottle of water, and he pours the water over the sand and the gravel and the rocks until the pitcher can hold no more.

"Now," he says, "the pitcher is full!"

The audience seems relieved.

"Someone here, tell me the moral of this story," he challenges them.

An eager young woman raises her hand.

"No matter how busy you get, you can always take on more!" she offers.

The speaker laughs and shakes his head.

"No," he says. "The moral of this story is to remember always to put the big rocks in first!"

Work v. Life

What comes to mind for many people when they hear that story is that life is a big rock and everything else is gravel and sand and, while it sounds nice to put big rocks ahead of the smaller stuff, it often doesn't turn out that way.

It's all about that phrase we hear so often these days: "Work/life balance." But unfortunately, work/life balance seems always to suggest that, while life is better than work, work always wins and life always loses. The goal is to try not to cheat life too badly in the "balance." In other words, we're always trying to steal time for life from work.

Now there's a sad statement, eh?

Most approaches to work/life balance tell us how to work smarter so we'll have more life because we're efficient enough to get work out of the way first. Some approaches to the balance question include mindfulness among their practices. By consciously remembering the importance of "being in the moment," we get more out of life, maximizing what little of it we have once we get off work.

Think about it: Which is better, life or work? Which is more important? It's easy to say, "Well, work is just a job, just a paycheck." But isn't there really much more to it than that?

Consider a different answer, that perhaps neither life nor work has an exclusive on "big rocks." Rather than feeling all the time that life is tugging at your heartstrings while work dominates your day, maybe it is the case that work and life exist on a continuum and are not at odds with each other after all.

We've lost sight in recent decades of the importance of work, how personally significant it is to most of us. People want there to be some meaning in the work they do, something more than hours of labor that result in a bi-weekly paycheck. The paycheck is necessary, of course, but at the end of the day, we want to think we've done something more than that.

In 1972, an extraordinary book was published that is, to this day, the definitive examination of people in jobs: *Working,* by world class journalist Studs Terkel.

He met and listened to workers around the nation, capturing their thoughts and feelings about the jobs they were doing, the on-the-job experiences they'd had. A recurring theme in this narrative is that employees want to feel they're doing something that matters.

"You throw yourself into things because you feel that important questions—self-discipline, goals, a meaning of your life—are carried out in your *work*," says one of the more than 100 workers whose voices Terkel shares.[1]

Keeping an eye on why our companies are in the business they're in—and what it means to the people who work there—sounds easy enough, but it isn't often enough what employees are reminded of.

Why is that? For one thing, it's difficult to measure purpose. There aren't generally accepted accounting practices that quantify how much of a company's operations are fulfilling that purpose.

Perhaps that's why it goes in and out of focus for so many employees. They're more often aware of the stock price, the latest cost-cutting measures, and the current steps being taken to meet that all-important end-of-quarter earnings report.

"You cannot inspire employees by urging them to help management get the company's stock price up," says Bill George in his book *Authentic Leadership.*[2]

"Employees today are seeking meaning in their work." We want to do our jobs because we *want* to do our jobs, to practice the professions we chose for ourselves years before, worked towards and attained. While the stock price matters to the company's health, it's tangential to employees' day-to-day lives.

Symptoms

A work/life balance tug-of-war sounds like this:

- ✔ *"It's just a paycheck. Why should I care?"*
- ✔ *"What I care about, I leave at home—which is where I should be right now!"*

- ✔ *"Considering all the lunches I miss, I should be at my goal weight by now! I give up too much for this place!"*

Causes

✔ **We like conflict.** No matter what we say about how focused we are on resolving—even avoiding!—conflict, the fact is we crave it. Even those of us who describe ourselves as "conflict avoidant"—that is, we don't like to engage in an active dispute—still like to segregate complex problems into two opposing camps. Setting up the conflict helps to clarify the problem, makes it easier for us to wrap our imagination around the issue. So we think of work and life as engaged in a heated contest for our time because it helps us to have clarity about the subject. Conflict creates clarity!

In this corner, we have **LIFE**

And in this corner, we have **WORK**

In a way, we've become conditioned to seeing work and life as contenders for our time and attention.

✔ **Work is prestigious; life is not.** In professional settings, we compete for positions, raises, plum assignments, recognition. When we earn them, it's a validation of our achievement. Some might go so far as to say it's a validation of their worth.

So work is where we demonstrate our accomplishments. That makes it special, something that's uniquely ours. Life, on the other hand, is something anyone can have. If you lose your job, you still have your mortgage, your dog, the lawn to mow, the dishwasher to empty.

Life is ordinary. Work has status.

Putting work on a pedestal is, to some extent, something we do ourselves. But don't let work make you feel more important than life makes you feel. Sometimes you'll have to remind yourself to elevate life to at least the same stature as work. Remember that life is made up of some priceless elements—your family, your health, your friends.

Recommendations

Know—and be true to—your boundaries. Remember you are a whole person, a person who has both a professional mode and a personal mode. Honor them both. There's no real Return on Investment (ROI) on obsessing about work or working long hours habitually because, after awhile, you work more and more slowly and with less and less inspiration. (See "Efficiency Opportunity" about taking the best advantage of your best hours!) When you don't shift out of your professional mode to your personal mode, you compromise yourself, your job performance, your company's purpose, your home life.

Remember the <u>purpose</u> of the business you work for. If you feel a **conflict** between work and life, that sinking feeling that life is always losing out to work, remember that every business has a purpose. To supply food. To deliver electricity. To design buildings. To conduct research. To build bridges across cultures. To renovate civic infrastructure. To promote learning, restore rivers, capture history, heal the sick, enable new businesses. That's what we *do*. We also make money—very important— but when it's the most important thing, or (worse) it's the *only* important thing, when we forget why we do what we do, we lose our sense of contribution to something greater.

In his *Harvard Business Review* article (December 2002), "What's a Business For?" Charles Handy recognizes the same phenomenon we talked about in the Overview: "The contribution ethic has always been a strong motivating force. To survive, even to prosper, is not enough. . . . We need to associate with a cause in order to give purpose to our lives."[3]

You might even **try having a conversation with your supervisor or manager about purpose.** "Hey, I read this book and it had some stuff about how we should all remember the purposes we're serving in our jobs . . . it was pretty good . . . wanna borrow it?" If by some chance your manager says "no" to that question (I can't imagine he or she would, but you never know!), you could quickly flip open this book and summarize the key points. Maybe your manager, too, has forgotten why he or she is there.

Whatever our company's purpose and our particular role within it, we need to make sure that it's more often front and center in our hearts, minds and conscious attention. Losing sight of our contribution to progress and improvements is what leaves us thinking work is Darth Vader and life is Luke Skywalker.

Personal Plan

Time again to turn to your **Personal Plan** in the Appendix section of this book and write specific tips and improvements that you want to remember and use in the future. Consider answering these questions in your **Personal Plan:**

What's the purpose of your company? Of your job within your company?

What did you always want to be when you "grew up"?

Are you doing that job, or something close to it, today?

Can you re-focus yourself on your talents and contribution?

Are you using boundaries effectively, knowing when you've spent enough time on something that's draining you?

Will you re-define boundaries for yourself in the future?

■○■○■○■

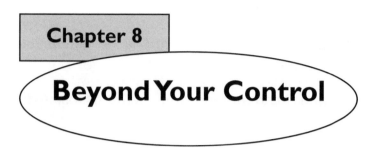

Overview

We've mostly been talking about things that are pretty much directly in your control, things you can affect and improve by simple, sometimes systematic, changes you personally can undertake. You might argue that work/life balance offers limited opportunities for you to do that, and I'd agree. Nonetheless, you can alter your outlook, and that's your choice, so to some extent you can make changes in that area.

It would be unrealistically cheery to suggest that you can control all the factors involved in managing many priorities. That's not all there is to it.

There are, of course, forces at work that complicate your "to do" list that are beyond your control. They come at you unexpectedly, and you must respond to them.

Today's workplaces are driven by competition, customer satisfaction and bottom-line concerns. The focus is often on containing cost, and this shows up in your every day life as change—new cost-cutting, belt-tightening measures, new procedures, new products, new or changing regulations, changing players, or even a changing of the guard, all of which bring you new priorities, new assignments, things which are beyond your control to anticipate, much less manage.

There's certainly an inherent conflict between change and control.

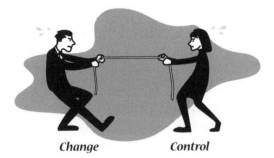

Change   *Control*

Change always wants things to be different, new, improved. Control always wants things the same, predictable, stable. (Remember: We love conflict, because it makes things clear!)

So while we may want control over our lives at work, the reality is that some are beyond our control.

However, even given all that, you're not a completely hapless victim. First of all, remember that everything we've talked about so far—Procrastination, Professionalism, Efficiency

Opportunities, Communications, and even to some extent Work/Life Balance—are areas where you, personally and directly, can effect change to help you make sense of—and manage—your many priorities. What we're talking about now— things "Beyond Your Control"—are areas where you can still apply initiative, influence and even optimism to improve your lot.

Symptoms

Those areas that are pretty far beyond your ability to control them directly look like this:

- **Demanding clients/customers**—Especially if there's any bad history or distrust among clients and customers, this can be a factor. Clients may not even have had the bad history with you or your company! They just have "baggage," as pop psychologists say, emotional history somewhere else that makes them critical and insistent.

- **"Do it NOW!"**—It's rare that anyone says "Do it whenever you feel like it, I'm in no hurry. . . . " "*Now!*" is more like what we usually hear. Of course some things really are due in the "Now!" timeframe— responding to outages and other emergencies, for example, but seeing that people can operate well, sometimes even exceptionally well, in emergency mode makes clients and managers ask for that kind of performance all the time!

✔ **Ineffective meetings and meetings you don't need to be in**—are another of the **Productivity Pillagers,** and the bane of life at work. They can take the form of meetings that don't get the job done, wander off topic, or require attendance on the part of people who really aren't needed there. Yet they still have to sit there for the entire hour . . . or two . . . or three. . . .

✔ **Meetings you wish you could go to but can't**—Meanwhile, while you're in a meeting you don't need to be in, there's another going on near by (a knowledge transfer session, a working design session, a conference about a problem) that you'd like to go to, but you can't because you're double-booked at that same time!

Causes

✔ **Work environments are dynamic,** priorities heap and shift, things change! Priority calls are tough to make. Businesses sometimes have to respond by shifting their priorities on the fly, which is frustrating for workers, especially if they don't know what's driving it—like competition or financial pressures, for example.

As cost-cutting continues to be all the rage for businesses large and small, one bit of fall-out is that every work effort claims to be a "top priority." That way, it's in a better position to compete for limited resources (people, money, technology, etc.). Politics like this can create enough momentum to propel a project into top-ranking priority status for awhile, only to be unseated by another project that has done the same thing. We may not particularly like that it works this way, but in many places, it is so.

✔ **Business is paradoxical!** Business leadership is conflicted because at the core of their planning, strategy and performance are fundamentally contradictory drivers. On the one hand, we say "Quality matters!" and in the next breath, we hear "Cut corners to save money!" But doesn't quality cost money? Similarly, we hear over and over the importance of short-term quarterly gains. But whatever you do, don't take your eye off the future, the strategy. Remember, too, that keeping costs down is of paramount importance, but so is growing your business. Can you "cut" and grow? Probably not, but you'd better be doing both.

✔ **There is no "One Right Answer"**—It's tempting to think that all this would be cleared up if only management would just stop changing its mind, the reality is that management has to keep changing its mind to respond to the paradoxical drivers of business as well as fluid conditions in the market, in its workforce, in the economy.

Recommendations

For **dynamic work environments** where you have to cope with the demands of many high-priority projects and work initiatives, there are a few personal strategies you can adopt:

Try "Reverse" time-boxing—Instead of promising yourself to work no less than 30 minutes on something (as when you're trying to beat procrastination), instead promise yourself to work no *more* than 30 minutes on something. That assures you of making some progress and still springing you at the end of 30 minutes to go work on something else. Some progress on several work efforts (instead of no progress on any) is more or less assured that way.

Business is paradoxical, and so people who insist on black-and-white solutions, will struggle with this one. For starters, the best thing you can do is to acknowledge that living with paradox is a workplace reality. You, personally, may not have to think about the implications of this every day (and it might make you crazy if you try to!). It's more likely the case that senior management gets to struggle with paradox directly. But recognizing that there is a variety of powerful influences at work here will help you accept that there are things you can't control without feeling like a victim.

Your "to do" list will continue to fluctuate because **there is no "One Right Answer."** If you're a purist, and you like it when decisions are clear and uncluttered, you can expect to be uncomfortable about this—uncomfortable, but not unhelpful.

Applying Emotional Intelligence

 The best thing you can do when forces beyond your control are making a mess of your otherwise orderly world is to apply some specific Emotional Intelligence (EI) capabilities to the situation. There are three specific EI competencies that can be used here:

- Initiative

- Influence

- Optimism

Applying Initiative, according to EI expert Daniel Goleman, means that you:

- **Seize opportunity:** Identify something that needs to be done, improved, clarified, and take it on;

- **Go beyond what's required:** That may mean doing it even if it isn't in your job description;

- **Mobilize others, are enterprising:** Getting others onboard with an idea or initiative.[1]

Here are "Beyond Your Control" symptoms and how applying initiative helps:

Symptom	EI to Apply: Initiative
Everything is a #1 priority. What's really top priority?	*"Here's what I think is the order of importance. Shall I use this to know what to tackle first, second, etc.?" You may be filling a void. If you're wrong about what's important, someone will tell you.*
Inefficient and/or ineffective meetings.	*"How about if I take meeting minutes and publish the action items to everyone?" You'll be happily surprised to discover how much you personally can control and improve a meeting this way.*

Daniel Goleman describes Optimism in EI terms as the ability to:

- Persist despite obstacles and setbacks;
- Operate from hope of success (not fear of failure);
- See setbacks as arising from circumstance (not personal failure).[2]

So here are some "Beyond Your Control" symptoms and how you can apply Optimism to improve things:

Symptom	EI to Apply: Optimism
Unreasonable customers'/clients' expectations.	*"I'll see customer situations from their perspective, not as criticism of me, and I'll rally to meet them. Eventually they'll trust us enough to relax."*
Things are always changing here.	*"Since change is part of business, I'll roll with it; I'll look for what should be better as a result of change."*

We hear a lot about influence and the importance of it in the workplace. In EI terms, Influence is about the ability to:

• Win people over
• Appeal to the listener
• Build consensus and support

Some ways to use your Influence to help manage many priorities when things are Beyond Your Control:

Symptom	EI to Apply: Influence
I'm not allowed to say "no" to my manager.	*"I'll use the credibility I've established with my manager to discuss this assignment realistically."*
Inefficient and/or ineffective meetings.	*"I'll talk to a meeting organizer I know and recommend improvements to format, sequence, timeframe, etc."*

Personal Plan Update

It's likely some of these "beyond your control" circumstances ring true for you. Take out your **Personal Plan** one more time, and write down what specific improvements and recommendations you plan to make.

Will you work on your personal credibility to bolster your ability to influence others?

Will you volunteer to help steer meetings in the hope of making them more efficient?

Will you take stock of how your manager views you (so you'll know whether you can say "no" now and then)?

Will you consider the perspective of the other person before taking action?

Will you take initiative, where otherwise you might hang back?

■○■○■○■

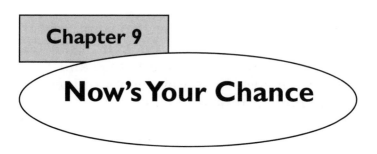

Chapter 9

Now's Your Chance

Before you congratulate your-self on having read a book this week, and before you close the cover on this volume and park it on your shelf for all eternity, ask yourself one question: **Why did you read this book?**

"I'm hopelessly disorganized and decided that reading a book about getting things done might help."

"I've read other books about 'time manage-ment,' but was looking for some new and dif-ferent ideas."

"I decided not to give up on the idea that I might actually get it together some day."

"My boss made me. It's one of my performance objectives, and it'll be on my annual review."

"I rented a vacation house and got snowed in. This was the only book on the shelf."

Whatever brought you to these pages in the first place, I hope that sometime since you started reading you've had at least one forehead-slapping moment when you said to yourself, "Ah ha! I could do that!"

That's really all you need, just one bit of inspiration you, personally, can adopt. What's yours? Are you going to:

✔ Try time-boxing?

✔ Take one bite?

✔ Perk up your status report?

✔ Adopt **Task Yourself**?

✔ Take more initiative?

✔ Clean out your email?

✔ Set goals for your writing?

✔ Avoid put-it-there-for-now syndrome?

Pick one or two that, when you read them, you knew instinctively "This is for me! I need this!" and try them for a couple of weeks. Be conscious that you're doing something differently. Then, in a week or two, take a breath and consider how, or whether, whatever change you've made has been helpful. It'll take you less than five minutes to size up whether it's working.

When you decide this has been a change for the better, pull this book down off the shelf again and cruise through your **Personal Plan** to find the specific tips you'll try next. Add one or two ideas every week or so.

You've invested the time, energy and imagination in examining the problem and a variety of solutions because something in you said you need to. You don't have to conduct a complete overhaul using everything you've read. Just try one tip at a time.

Now's your chance. You can do this.

■○■○■○■

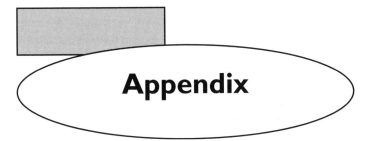

Appendix

Ask Yourself

Ask Yourself Score Sheet

Your Personal Plan

Task Yourself

Ask Yourself

Read each statement, the put a check mark in the column under the heading that best describes your reaction to the statement.

	Strongly Agree	Agree	Neutral	Disagree	Strongly Disagree
When I think something is hard to do, I tend to put it off until later.					
I do my best work under the pressure of a rapidly approaching deadline.					
I often underestimate how long it will take me to do something.					
I am often late to meetings.					
I often return phone calls later than what's considered an acceptable time.					
I think of myself as someone who is relatively disorganized compared to others with whom I work.					
I have no idea what my most productive time is during the day.					
I thoroughly read every email I get every day.					
When I get into disputes at work, I let it get to me; I dwell on the interpersonal conflict, or what's on the rumor mill about disputes.					

Ask Yourself — cont.

I am a perfectionist. I like everything done right, by my standards of "right."					
I am often worried or anxious about things at work.					
I don't ask for assistance from others at work very often.					
I rarely, or never, tell my boss when I'm swamped or overloaded.					
I generally say "Everything is under control" when I don't think it is.					
Don't ask me to give you "status" or "updates" when I'm busy! It takes too long.					
I talk about stress a lot, especially the stress I personally feel.					
I get interrupted a lot while I'm at work, and it sets me back.					
I have trouble saying "No" when someone asks me to do something if I know it's going to be difficult for me to get it done.					
I regularly give up personal time to get my job done, and I resent it.					

Ask Yourself — cont.

I would exercise more often if I could find the time.					
I have a hobby or pastime I enjoy, but I rarely do it anymore.					
I often skip lunch or eat at my desk so I can use the extra time to work.					
I often have to be many places at once. I'm needed here, there, somewhere else, all at the same time!					
I'm not allowed to say "No" to my manager when he/she asks me to do something.					
Everything around here is a #1 priority. We need direction and agreement about what's really a top priority and what can wait if it has to.					
Our customers/ clients are unreasonable in their expectations about when things need to be done.					
Ours is a fast-paced environment. No time to figure it out—get it done NOW!					
Things are always changing here. If they would just stay put for awhile, I'd get more done.					
I waste a lot of time in inefficient and/or ineffective meetings.					

Ask Yourself Score Sheet

Write the accumulated score for this section in the TOTAL row. If your score for this section is in the 3–9 range, this is an area of improvement for you.

	Strongly Agree	Agree	Neutral	Disagree	Strongly Disagree
P R O C R A S T I N A T I O N					
When I think something is hard to do, I tend to put it off until later.	1	2	3	4	5
I do my best work under the pressure of a rapidly approaching deadline.	1	2	3	4	5
I often underestimate how long it will take me to do something.	1	2	3	4	5
TOTAL for PROCRASTINATION					_____

Write the accumulated score for this section in the TOTAL row. If your score for this section is in the 3–9 range, this is an area of improvement for you.

	Strongly Agree	Agree	Neutral	Disagree	Strongly Disagree
P R O F E S S I O N A L I S M					
I am often late to meetings.	1	2	3	4	5
I often return phone calls later than what's considered an acceptable time.	1	2	3	4	5
I think of myself as someone who is relatively disorganized compared to others with whom I work.	1	2	3	4	5
TOTAL for PROFESSIONALISM					_____

Write the accumulated score for this section in the TOTAL row. If your score for this section is in the 5–15 range, this is an area of improvement for you.

	Strongly Agree	Agree	Neutral	Disagree	Strongly Disagree
EFFICIENCY OPPORTUNITIES					
I have no idea what my most productive time is during the day.	1	2	3	4	5
I thoroughly read every email I get every day.	1	2	3	4	5
When I get into disputes at work, I let it get to me; I dwell on the interpersonal conflict, or what's on the rumor mill about disputes.	1	2	3	4	5
I am a perfectionist. I like everything done right, by my standards of "right."	1	2	3	4	5
I am often worried or anxious about things at work.	1	2	3	4	5
TOTAL for EFFICIENCY OPPORTUNITIES				_____	

Write the accumulated score for this section in the TOTAL row. If your score for this section is in the 7–21 range, this is an area of improvement for you.

	Strongly Agree	Agree	Neutral	Disagree	Strongly Disagree
C O M M U N I C A T I O N S					
I don't ask for assistance from others at work very often.	1	2	3	4	5
I rarely, or never, tell my boss when I'm swamped or overloaded.	1	2	3	4	5
I generally say "Everything is under control" when I don't think it is.	1	2	3	4	5
Don't ask me to give you a "status" or "updates" when I'm busy. It takes too long!	1	2	3	4	5
I talk about stress a lot, especially the stress I personally feel.	1	2	3	4	5
I get interrupted a lot while I'm at work, and it sets me back.	1	2	3	4	5
I have trouble saying "no" when someone asks me to do something if I know it's going to be difficult for me to get it done.	1	2	3	4	5
TOTAL for COMMUNICATIONS					_____

Write the accumulated score for this section in the TOTAL row. If your score for this section is in the 4–12 range, this is an area of improvement for you.

	Strongly Agree	Agree	Neutral	Disagree	Strongly Disagree
W O R K / L I F E B A L A N C E					
I regularly give up personal time to get my job done.	1	2	3	4	5
I would exercise more often if I could find the time.	1	2	3	4	5
I have a hobby or pastime I enjoy, but I rarely get time to do it anymore.	1	2	3	4	5
I often skip lunch or eat at my desk so I can use the extra time to work.	1	2	3	4	5
TOTAL for WORK/LIFE BALANCE					_____

Write the accumulated score for this section in the TOTAL row. If your score for this section is in the 7–21 range, this is an area of improvement for you.

	Strongly Agree	Agree	Neutral	Disagree	Strongly Disagree
B E Y O N D Y O U R C O N T R O L					
I often have to be many places at once. I'm needed over here and over there and also somewhere else—all at the same time!	1	2	3	4	5
I'm not allowed to say "no" to my manager when he/she asks me to do something.	1	2	3	4	5
Everything around here is a #1 priority. We need direction and agreement about what's really top priority and what can wait if it has to.	1	2	3	4	5
Our customers/ clients are unreasonable in their expectations about when things need to be done.	1	2	3	4	5
Ours is a fast-paced place, no time to figure this all out. Get it done "NOW!"	1	2	3	4	5
Things are always changing here. If things would just "stay put" for awhile, I'd get more done.	1	2	3	4	5
I waste a lot of time in inefficient and/or ineffective meetings.	1	2	3	4	5
TOTAL for this section					_____

Your Personal Plan

For:	
Date:	

PROCRASTINATION	1	
	2	
	3	
	4	
	5	
	6	
PROFESSIONALISM	1	
	2	
	3	
	4	
	5	
	6	

Your Personal Plan — cont.

EFFICIENCY OPPORTUNITY	*1*
	2
	3
	4
	5
	6
COMMUNICATIONS	*1*
	2
	3
	4
	5
	6

Your Personal Plan — cont.

WORK / LIFE BALANCE	*1*	
	2	
	3	
	4	
	5	
	6	
BEYOND YOUR CONTROL	*1*	
	2	
	3	
	4	
	5	
	6	

Task Yourself

Priority	Sequence	Project	Task	Due	Notes
a	1	Alpha	Test	1/6	Resume
b	1	Wichita	Fix line down	1/9	Call vendor (800) 511-1111
a	2	Work Tracking	Product evals	1/29	
c	1		Call Dr. for appt.	2/15	Ofc moved— need phone #
B	2		Write up self eval	1/15	
B	3		Teacher conf appt	1/20	Set appointment
A	3	Work Tracking	Prep for meeting	1/7	Need ppt slides from Alice

This simple spreadsheet can serve to track pending tasks.

Priority = What overall importance does this task have?

   a = highest visibility, soonest deadline, most critical

   b = important, must keep an eye on, coming up next

   c = in the future, can't lose sight of, not now but later, mustn't drop it

Sequence = Within priority, which should be tackled first, second, third, etc.

Project = If there is a specific project associated with this task, list it here. If not, like a family occasion, leave it blank.

Task = The description of the task itself.

Due = The date it's due.

Notes = Any miscellaneous information you want to include that's relevant to the task.

You can add columns to this spreadsheet to make it more comprehensive (or more complicated). For example:

Dependencies—What tasks must be completed before this one can begin. You can identify a predecessor with a combination of "Priority" and "Sequence." For example, you can list simply **b1.**

Start Date—When does (or should) a task begin?

Status—What state is this task in? Pending? Completed? Approved? If it's important to you to track the progress of a task's status, this might be worth adding.

One final note: The more detail you add to **Task Yourself,** the more there is to maintain. Think about what's important to you to stay on top of. Anything else, omit.

Chapter 3 / Procrastination

1. Dr. Timothy Pychyl, Procrastination Research Group Home Page (www.carleton.ca/~tpychyl)

2. Daniel Goleman, *Working With Emotional Intelligence* (New York: Bantam Books, 1998), 318.

Chapter 4 / Professionalism

1. Wikipedia on "Professionalism," http://en.wikipedia.org/wiki/Professional accessed October 15, 2005).

2. Goleman, 93.

3. Richard Farson, *Management of the Absurd* (New York: Simon and Schuster, 1996), 130.

4. Goleman, 90.

Chapter 5 / Efficiency Opportunities

1. Jeff Davidson, *The Complete Idiot's Guide to Getting Things Done* (New York: The Penguin Group, 2005), 46.

2. Goleman, 82.

Chapter 7 / Work/Life Balance

1. Studs Terkel, *Working* (New York: The New Press, 1972), 521.

2. Bill George, *Authentic Leadership* (San Francisco: Jossey-Bass, 2003), 64.

3. Charles Handy, "What's a Business For?" in *Harvard Business Review on Corporate Responsibility* ((Boston: Harvard Business School Publishing Corporation, 2003), 81.

Chapter 8 / Beyond Your Control

1. Goleman, 122.

Telephone Orders: Call 503.635.0005

Email Orders: Orders@Auxilium-Inc.com

Postal Orders: Alder Business Publishing,
15800 SW Boones Ferry Rd., Suite A3, Lake
Oswego, Oregon 97035.

Fax: 503.293.8499

Item	Quantity	Amount	Item Total
You *Can't* Manage Time		$15.95	$
Shipping and Handling (first book, U.S. only)		$ 3.95	$
Shipping and Handling (for each additional book to the same address)		$1.95	$
Shipping and Handling (for each additional address)		$3.95	$
Note: No Sales Tax (state of origin has no Sales Tax)	N/A	$.00	$0.00
Order Total			$

I understand I may return items for a full refund for any reason.

Name: _____

Ship-to Address: _____

Billing Address (if different): _____

City: _____ **State:** ____ **Zip:** _____

Email address: _____

To contact the author directly:

503.387.5959

susan@susandelavergne.com

Website:

www.susandelavergne.com